A CANALSIDE CAMERA 1845-1930

Michael E. Ware FRPS

David & Charles
Newton Abbot · London
North Pomfret (Vt) · Vancouver

Dedicated to the late Tom Rolt – who
gave me my interest in the British
canals through his book *Narrow Boat*.

ISBN 0 7153 7001 4
Library of Congress Catalog Card Number 75-2916

© Michael E. Ware 1975

Set in 10 on 12 Erhardt
and printed in Great Britain
by Jolly & Barber Ltd, Rugby
for David & Charles (Publishers) Limited
Brunel House Newton Abbot Devon

Published in the United States of America
by David & Charles Inc
North Pomfret Vermont 05053 USA

Published in Canada
by Douglas David & Charles Limited
132 Philip Avenue North Vancouver BC

A CANALSIDE
CAMERA 1845-1930

Taken in 1916, this is a typical canalside
scene at Ironbridge Lock in Cassiobury
Park on the Braunston to London
section of the Grand Junction Canal.
The boat horse is about to take the
strain and pull the wide boat from the
lock; a pulley block at the masthead
eases the initial weight. In practice, the
14ft 3in wide locks allowed two full-
length narrow boats to pass through a
lock at the same time but some carriers,
notably Fellows, Morton & Clayton,
developed a wider type of boat for use
on this section.

CONTENTS

INTRODUCTION

The years between 1845 and 1930 saw the heyday of the British canal system. Photography had not been invented when the first canals were being built but even during the mid-Victorian period the new medium was rarely used to record industry and transport. Thus, there are few really good photographs with which to tell the early canalside story. By the turn of the century photographers had become more adventurous and a truer picture of the colour and drama of canal life of the period that followed survives. After 1930 the winds of change began to blow, the Grand Union Canal was being modernised, the horse-drawn boat had nearly disappeared in favour of motor-driven craft and much of the bustle had gone as the roads and railways lured the last of the cargoes away from the waterways.

Today it is commonplace for an industrial company to employ a full-time photographer but highly unlikely before 1930. Thomas William Millner of the Grand Junction Canal Company did, however, often carry his camera in the course of his duties as assistant engineer. Three books of his photographs record much of the repair and building work on the canal over a period of twenty years and are now kept in the archives of the Waterways Museum at Stoke Bruerne. Some of Millner's photographs have been used here and although they all depict the Grand Junction the subjects chosen for inclusion could have applied to almost any man-made waterway in Britain.

In medieval times the rivers were the only reasonable means of transport. The Romans had left a road system behind them which the British had failed to maintain and the movement of freight across the country was slow, laborious and sometimes even impossible. River traffic was fraught with difficulty and it was not until

1   T. W. Millner was appointed assistant engineer to the Grand Junction Canal Company's northern section around 1892. In 1916 he took on extra responsibilities and continued in office right through until the days of the Grand Union Canal Company in the early 1930s. His personal transport during part of this period was a Bradbury motor cycle, though in 1918 the company bought him a secondhand Belsize car for £185. He is sitting astride the faithful Bradbury, the product of a firm which started making motor cycles in 1902 in Oldham and which ceased production in 1925. This is a fully equipped example, having acetylene lighting, speedometer and a horn as well as an unusual feature for a motor cycle – an exhaust whistle mounted down by the engine. During his period of office he took many photographs of repairs and alterations to the Canal; his quarterplate wooden camera and tripod, along with his photographs, are preserved at the Waterways Museum, Stoke Bruerne.

the artificial waterway became a reality that water transport came into its own.

The majority of Britain's canals are unique on account of the very small dimensions to which they were built. They were designed at a time when labour was cheap but, when other forms of transport took away their traffic, the canal companies had no money to modernise in order to compete.

In England the first artificial inland waterway to be constructed was the Sankey Canal near St Helens, built by Henry Berry of Liverpool in 1757. The first canal in the British Isles was the Newry Navigation in Northern Ireland which opened in March 1742 but the third Duke of Bridgewater, with his agent John Gilbert and engineer James Brindley, usually received the credit for the conception of the British canal. Their canal, the Bridgewater, was completed in 1761 and ran from Worsley to Manchester. This 10½-mile 'cut', which included an aqueduct over the river Irwell but had no locks, made an almost overnight reduction in the price of coal in Manchester from 7d to 4d a hundredweight. Coal was the staple diet of the Industrial Revolution. A cheap transport system was needed to get it to the industrial areas and, just as important, to take away the manufactured goods. It was for this reason that Josiah Wedgwood approached Brindley to survey the line of a canal that would link the Trent with the Mersey and later join this to one linking the Trent with the Severn.

Wedgwood had established his pottery at Etruria near Stoke-on-Trent in order to be near the coalfields. His clay came from Bovey Tracey in Devon and was transported overland at great expense from coastal vessels docking on the River Weaver. His finished pottery was again taken overland to the River Weaver for

2 The Thames, which is now navigable as far as Lechlade, has not always been in such good condition; lock-building began in the 1620s and was only completed in the present century. Well into the canal age there were complaints of insufficient depth on the upper parts of the river and traders coming from the Thames & Severn and the Wilts & Berks Canals had many difficulties in getting through to the lower reaches.

These Western barges lying at Windsor Bridge would normally have traded as far as Oxford, though sometimes going on to the higher reaches. They were photographed by William Henry Fox Talbot in 1845, making this one of the earliest photographs of this country's navigable waterways in existence. It was exposed only ten years after Fox Talbot took Britain's first ever photograph at Lacock Abbey.

3   The River Ouse at York has long been used for trading; in 1757 Naburn lock five miles below York was built, thus finely completing a reasonable navigation from the River Humber. Though the river is not tidal in York, Bradshaw records that it is liable to flood as much as 12 ft and on one occasion 16ft 6in above normal summer level. This historic photograph was taken by Roger Fenton in 1854, a year before he left Britain to be the first ever war photographer with his vivid coverage of the Crimean campaign. The scene is Lendal Ferry in York, with trading keels moored on the right whilst the foreground is occupied by what appear to be houseboats.

shipment to British coastal towns and abroad. The breakages were enormous, one source quoting them as 90 per cent. The Trent & Mersey Canal was finally opened in 1777.

The success of the early waterways then snowballed and 'canal mania' set in. It seemed that every town and city wanted a canal and eventually the country had a great network of such waterways, the last one of any importance to be built being the Birmingham & Liverpool Junction Canal in 1835, later to become the Shropshire Union Canal mainline. Ironically, it was on a tramway connected to the Glamorganshire Canal that Trevithick tested the first railway locomotive in 1804. In 1825 the Stockton & Darlington Railway opened, followed five years later by the Liverpool & Manchester Railway and by 1845 'railway mania' was developing. The canals now began to feel the effect of this new form of transportation, since the earlier railways had at first concentrated on passenger traffic rather than goods. Slowly the railways took away the cargoes from the waterways and the canal system went into decline. Some waterways closed and others sold out to their railway rivals, who usually stifled them, even if they let them stay open. On the improving road system, however, the steam engine did not have the same effect as it was virtually taxed off the turnpikes by penal toll charges and other legislation.

After the First World War, hundreds of ex-War Department lorries were brought back from France, overhauled and sold cheaply to many firms that were then springing up in the road haulage business. By the time the Second World War was over, many hundreds of miles of canal had been closed. In 1972 the last regular long-haul commercial cargo was carried on the smaller inland waterways of Britain. Working narrow boats are still to be seen trading today, but there is little regular work for them. Trade still happily exists on the navigable rivers such as the Humber, Ouse and Trent, as well as on the larger canals more suited to modern transport needs, such as the Gloucester & Sharpness, or the Manchester Ship Canal.

Though outside the scope of this book, the canals have now taken on a new lease of life for recreational purposes and slowly over the last thirty years, aided by the Inland Waterways Association and other interested parties, they have been adapting to their new role. Canal cruising is now big business, with over a hundred firms offering boats for hire, and there are long waiting lists at the boatbuilders for new craft, many of which are being built on traditional lines.

CONSTRUCTION

4   The building of locks involved the employment of many skilled workmen, such as masons, carpenters and bricklayers, whose tools and techniques changed little over the next 150 years. This photograph shows a lock under repair and although taken around 1900 little has changed over the years. The carpenters are working on the sill on the lock floor, against which the finished gate will fit. The labourers are scooping out the water trickling into the chamber with crude wooden scoops (or scopes). These scoops are also used for puddling the clay which lines the canal bed and for baling out boats. The paddle hole for the release of the water from the full lock can be clearly seen on the new gates awaiting fitting. In the Fens these paddles are often referred to as 'slackers' and in the north of England as 'cloughs'. A lock such as this passes on average 56,000 gallons of water each time it is emptied, which gives some idea of the pressure of water behind the lock gate. The work of the carpenter, therefore, has to be very accurate.

5   There was little mechanical aid
available to the navvies and contractors
who built the waterways in the 1700s
and 1800s. All the digging was done by
hand and the spoil carried away either
in a wheelbarrow or by horse and cart.
Building canals was dangerous work,
and death and injury were common.
During the rebuilding of the Watford
flight of locks between 1901 and 1903,
almost the same methods were
employed as had been used by the
pioneers, with men carrying spoil along
prepared wheelbarrow runs. On the
steeper ascents such runs became very
slippery and caused many accidents.

6   One of the most important aspects
of canal building was to ensure that the
waterway was watertight. The most
usual method was to prepare a mixture
of clay and water, called 'puddle', which
was spread thickly along the bed and
shoulders of the canal rather like a stiff
concrete. It is reputed that contractors
would herd cattle along newly puddled
lengths and the tread of their cloven
hooves would pack the puddle in well.
The Thames & Severn Canal on its
summit 'pound' (the length between
two locks) was beset with leakage
problems and had to be rebuilt several
times. In 1902, shortly after the Canal
had been taken over by the
Gloucestershire County Council, work
started again on this pound. This time,
as well as puddling, part of the length
was encased in concrete and a
temporary tramway laid along the
towpath to help with bringing in
materials.

7   Each stretch of canal cut across
farmland and country roads and tracks;
since the farmers needed access to their
fields, many bridges had to be built over
the canals. These were mainly of the
'hump-backed' style as the shorter the
approach gradient, the cheaper it was
for the canal company to build. The
steepness of the approach as well as the
required width of the carriageway was
often defined in the canal's Act. The
perfect shape of the brick arch was
obtained by building the bricks over a
wooden former, called the centring.
Traditional methods were used in 1904
to rebuild Hunton Bridge (162) on the
Grand Junction Canal near Abbots
Langley. The demolition of the old
bridge and the building of the new one
took two months and cost £601.

8   Mechanical aids were first used in waterway construction in Britain during the building of the Manchester Ship Canal. Huge mechanical excavators, such as this Ruston & Proctor steam navvy named Jumbo, could move as much as 2,500 tons of earth in a twelve-hour shift. The spoil was carried away on a network of 228 miles of temporary railway track. Horses were employed for the initial marshalling of the spoil trains which were then hauled away by locomotives. These trains in turn brought back coal for the steam navvies, which also required a continuous source of clean water for their boilers. 6,300 wagons were at work on this project, together with 196 horses, 173 locomotives and 97 steam excavators.

9

10

9 Much ingenuity was used to solve the problems of cutting the Manchester Ship Canal. In 1888, on the section from Ince to Weston Point, a specially designed German excavator was used. There were in all seven land-based excavators, each capable of moving up to 2,000 tons of earth a day; the buckets moved uphill on an endless belt and discharged into railway wagons which passed under the body of the excavator. It is a pity that this photograph cannot be compared with a scene in the 1700s, when hundreds of navvies aided only by hand tools, horses and carts would have been needed to make a cutting at this depth.

10 Building tunnels was a difficult and dangerous task for the canal navvies who often used techniques gained from coalmining experience. As well as digging in from each side of the hill, digging was undertaken outwards from the foot of vertical shafts sunk from the surface to the tunnel level. The men and spoil were winched to the surface up the shafts, either by horse gin or by hand, as in this case. Sometimes the spoil was taken away by horse and cart but more often it was just dumped around the head of the shaft, as at Blisworth Tunnel. After completion, the shaft would sometimes be bricked around and used for ventilation, although often, as at Harecastle Tunnel, the shafts were closed up. With the advent of motor craft, ventilation was sorely needed and because of the lack of shafts, a large fan has had to be used to suck the fumes through Harecastle Tunnel.

11 Canal navvies were a hard-living, hard-working, hard-drinking crowd of men, often of Irish descent, who moved around the country from job to job. The contractors who hired them gave them rudimentary shelter in communal huts but food and clothing were the responsibility of the men themselves. The pay was low; in 1770 a contractor on the Oxford Canal was paying a little over one shilling per day. Carpenters, bricklayers and blacksmiths, and men with mining experience for the tunnels were paid higher rates. In 1794 on the Lancaster Canal project, masons received three shillings per day, whilst the men cutting the stone in nearby quarries were paid two shillings and sixpence. Instead of cash the men were often paid with tokens which were only exchangeable in the contractors' own shops, where food and goods were available at inflated prices. After the run down of the 'canal mania', many navvies and their descendants moved on to building the new railways but by the end of the 1880s this source of work had almost ceased. These navvies were working on the Manchester Ship Canal in 1887 which provided employment for 700 men initially, rising at one stage to 1,700.

2 Child labour was also used during the construction of the canals. These four children are outside a temporary clay shelter.

13 The navvies lived in very poor huts, often made out of turf and sometimes they had to pay rent to the contractor or canal company for the privilege. By the time the Manchester Ship Canal was being built, conditions had improved, as shown by this navvy's house at Moore. Huts, meeting rooms, schools and hospitals were built by the contractor, T. A. Walker, for the workers. The huts were mainly on isolated lengths of the works and some navvy towns were constructed, such as Marshville on the Frodsham Marshes. Other workers found lodgings in Warrington, Runcorn and surrounding areas.

14 An enterprising local tradesman, thinking that there was money to be made from the workmen on the Manchester Ship Canal site, converted the hulk of an old sailing flat into a floating cafe and dosshouse. Hard liquor was the undoing of most navvies, although this establishment does not advertise the selling of intoxicants.

14

LOCKS, LIFTS, BRIDGES AND TUNNELS

15 The simplest form of lock is the flash lock or staunch. This consists of a single gate built into a weir which when opened allowed craft to pass, those travelling downstream being carried by the flow through the gap, those proceeding upstream having to be hauled against the current. The first flash locks on the River Thames were built in the thirteenth century and this one at Eynsham, one of the last locks of its type in the country, was only replaced by a pound lock in 1931. In this instance the river was in flood, making the passage by any boat very difficult. So strong was the current that the gate and bridge were weighted down with stones and tied back to prevent them being floated off. The weir paddles, normally operated by a simple windlass and chain, have been completely withdrawn so as not to impede the flood water.

16 Before the coming of canals, a number of rivers had been made navigable by the addition of weirs and locks. In 1721 an Act was passed with regard to the Rivers Mersey and Irwell so that they would become, in the words of the Act, 'very beneficial to Trade, advantageous to the poor, and convenient for the carriage of Coals, Cannel, Stone, Timber and other Goods, Wares and Merchandizes, to and from the Towns and parts adjacent . . .' Barton lock, photographed here in March 1888, was typical of a river pound lock. The disturbance to the land on the right was due to the impending works in building the Manchester Ship Canal.

17 The average size of a lock on the narrow canals is approximately 72ft long by 7ft wide. The lock in the foreground was, at that time, the bottom lock (No 5) of the Parkhead, Dudley, flight and No 4 is just visible through Dudley Lye Waste bridge. This flight was opened in 1792. By 1893 mining subsidence was affecting the bottom lock and the present Blowers Green Lock was built on the site of No 4 to replace them both. The new lock has a depth of 11ft 3in, deeper than the average 6 to 8 feet of most British canal locks. Industrial archaeologists will be particularly interested to see the beam engine and 'egg' boiler of the abandoned Peartree Colliery. The date of this picture could be any time between 1866 and 1893.

18 On the Shrewsbury Canal, opened in 1797, the lower gates on each of the locks were of the guillotine type. The single bottom gate was raised by means of a chain on a winding drum and counterbalanced by a large box of stones, hidden behind the gate in this picture. The operating gear of the single paddle can be clearly seen built into the gate, and emptying would have been speeded up by partially lifting the gate. The locks on this section of the canal were over 81ft long but only 6ft 7in wide and could pass four tub boats at a time. In later years the Shropshire Union had to build 24 special 6ft 2in beam narrow boats to work this piece of the canal. They worked as far afield as Ellesmere Port. A pair of guillotine gates remain in situ – though not working – at the junction of the Stratford-upon-Avon Canal and the Worcester & Birmingham Canal at Kings Norton junction. Originally these were built as a stop lock to ensure that there was no continuous flow of water from one canal to another as the water levels were almost identical. Such was the jealousy with which canal companies guarded their water supplies.

19 Where a canal had to be taken over a hillside, it was not always possible to build a flight of locks, perhaps because of insufficient space for building, or an inadequate water supply to keep them filled if the traffic was heavy. In some cases the canal engineers turned to the inclined plane as an alternative, usually used by tub boats which were nothing more than floating boxes. On arrival at the plane they were floated onto a cradle running on railway lines. In most cases, loaded tub boats descended, pulling up the empty ones, and the steam engine, which was housed in the winding house at the summit, was only used to pull them over the reverse slope at the top. This is the Trench inclined plane on the Shrewsbury Canal and was the last to work in this country, closing in 1921.

20 The Hay inclined plane on the Shropshire Canal was built to lower tub boats down to a short stretch of canal parallel to the River Severn at Coalport, at which point the cargo could be transhipped into trows trading on the river. This incline became fully operational in 1793 and was steam-powered; the winding house is just visible at the top. The incline was equivalent to approximately twenty-seven locks and could pass a pair of five-ton boats in only three and a half minutes. It was worked by four men. It had hard usage for the first sixty years of its life but by 1855 much of its connecting canal system had been affected by mining subsidence and trade slowly declined. The incline, which had a length of 350yd and a vertical fall of 270ft, was last used in 1894, about the time of this photograph.

21 In 1900 the skyline at the little village of Foxton, near Market Harborough, was changed by the building of the Foxton inclined plane. This massive structure was built as an alternative to the bottleneck of the ten Foxton locks which brought the Leicester section of the Grand Junction Canal down from its summit level by two staircases of five locks each. The inclined plane was also part of a scheme to provide a wide waterway to London for the Derbyshire coal traffic, as the plane could carry one barge or two boats at the same time. A similar plane was considered for the locks at the other end of the summit at Watford, but the anticipated coal traffic did not materialise and it was never built, leaving that flight of narrow locks as the only barrier for wide beam traffic. The Foxton locks were never closed; they rise to the right of the picture, the first lock being immediately under the bridge.

22 Craft were floated into the two caissons which ascended and descended sideways on sixteen wheels and eight rails on an incline 307ft long, with a total drop of 75ft and a gradient of 1 in 4. Clearly seen is the change of gradient at the top of the plane; the gradient of the main track eased to compensate for the loss of weight and counterbalancing effect when the lower caisson entered the water. Extra outer rails gave support to extra carrying wheels and brought the caisson absolutely level and aligned with the top arm of the canal. A double-cylinder high pressure steam engine was used to provide movement. The total cost of the installation was £39,224; though it could be worked by only three men it was found to be uneconomical only eight years after its opening. By 1910 it was only being used during the daytime and by 1912 most traffic was again using the narrow locks. It was dismantled in 1926 and sold for scrap two years later.

23   The only remaining operable canal lift in Britain is a vertical lift as opposed to the inclined planes previously illustrated. It is situated at Anderton in Cheshire and moves boats from the Trent & Mersey Canal down fifty feet to the River Weaver below. Planned by Edward Leader Williams (later Sir Edward) and designed by Edwin Clark, the lift was opened in 1875 and originally worked by counterbalanced caissons. In 1908 the Weaver Navigation staff converted it to run on electricity, each caisson being powered by an independent electric motor conterbalanced by weights. This photograph was taken shortly before conversion. There was much trade between the Canal and the River Weaver, on whose banks stood many chemical works requiring both coal and salt. Canal cargoes were also transferred to Weaver flats by lowering the goods down the staithes and chutes which flanked the great lift. Most of the Anderton traffic was destined for Liverpool or abroad, and raw materials for the Potteries came up the Weaver from Weston Point.

24   The Barton Swing Aqueduct was designed by Sir Edward Leader Williams to replace Brindley's aqueduct over the River Irwell on the Bridgewater Canal and is just as much an engineering feat as Brindley's original structure. It has a swinging span of 23ft, carrying a waterway 19ft wide and 6ft deep. When a ship passes along the Manchester Ship Canal, this 1,450-ton structure is swung aside by hydraulic power. There is also a hydraulic ram to lift the span vertically to take 900 tons of weight off the central roller-bearing race; otherwise friction would be excessive. It was not usual to swing the aqueduct with craft on it. Clearly shown are the raised towpath to allow maximum width at water level and the sealing rubbers on the end of the watertight section.

25

26

25 Sapperton Tunnel on the Thames & Severn Canal is the third longest tunnel on British waterways. It was completed in 1789 and was 3,817yds long. In order to keep construction costs down, there was no towing path through this and many other tunnels, so all boats had to be legged through – pushing with the feet against the tunnel walls and walking the boat through. The bylaws specifically mentioned that 'shafts, or sticks or other things against the arch of the tunnel was prohibited'. As boats could not pass in the tunnel, entry was at fixed times at either end, the journey taking between three and five hours. Another hazard was the risk of grounding in the tunnel itself, due to the lack of water on this section, and boats often had to partially unload their cargoes into lighters kept at each end of the tunnel for this purpose. This is the western end at the head of the Golden Valley; many of the castellations are missing nowadays.

26 During 1868 the first part of the 788yd tunnel at Fenny Compton on the lower Oxford Canal was opened out; it had originally been built in two sections with an opening in between. At no point could it have been more than 40ft below the surface, and so it may seem strange that it should have been a tunnel at all. Early canal engineers did not like cuttings because of the likelihood of landslips and the amount of spoil which would have to be removed at the time of building, and they preferred to make tunnels. Two good examples of these shallow tunnels are Barnton and Saltersford on the Trent & Mersey, where the Canal is carried on a terrace alongside the Weaver valley and a landslip would lead to the collapse of the Canal. The second section of the Fenny Compton tunnel was left where it was, crossed by a road, but in 1870 this too was opened out and a new bridge built. The small bore of the original tunnel can be clearly seen with a maintenance boat fitting it very closely.

27 A familiar sight on the British canal system is the humped-backed bridge, sometimes painted white. Where the canal narrowed under the arch, the canal builders often took the opportunity to construct grooves for 'stop planks' which could be inserted in case of emergency to seal off a section of the waterway. This narrowing is usually referred to as a 'bridgehole'. All bridges are numbered and on the Staffordshire & Worcestershire Canal, each bridge carries a name plate as well. The Grand Canal in Ireland has the bridge name and date of construction carved in the stonework. Sometimes bridges were used to change the towpath from one side to another, because of land purchase problems or to minimise embanking. This scene is on the Wirral Canal near Ellesmere Port in 1894.

28 The lock-keeper's house has taken many forms, but none so unusual as those on the Thames & Severn Canal which ran from a junction with the River Thames above Lechlade through to the Stroudwater Navigation at Wallbridge, Stroud. On this Canal some of the houses were round and shaped like a watch tower, to enable the keeper or lengthsman to see approaching craft. This is the round house at Inglesham where the canal joins the Thames. On the left is the towpath bridge over the river and the Canal Company's warehouse. The lower floor of the round house was also used as a stable. This photograph dates from 1918, by which time most of the trade had ceased on the Canal. There were also a few round houses built on the Staffordshire & Worcestershire canal, the one at Gailey lock being the best example.

29 Besides the familiar stone or brick built bridges, there were many varieties of lifting bridge. This type of bridge was built where there was unlikely to be much traffic, probably connecting two fields rather than carrying a road or track. This variety of 'accommodation' bridge on the lower reaches of the Oxford Canal is a typical example. At the time of this photograph in 1868, bridge 241 was in open country but now the bridge has gone and houses replace the fields. The simple, almost crude method of using roughly shaped tree trunks as balance beams can be clearly seen.

30 The pumping station was a very important feature of the canal landscape. It was often necessary to pump water up from a lower level in order to keep a full pound. This water could come from a specially constructed reservoir or a nearby river or stream. It might even be pumped back from the bottom of a flight of locks, such as at Braunston on the Grand Junction Canal. The Braunston steam pump was installed in 1805 and the pumping station rebuilt, as seen here in 1900, to take a single-crank tandem compound engine, driving a centrifugal pump. The method of construction clearly shows how all the machinery was first installed and then the building put up around it. In turn, this pump was superseded by a diesel in 1940 and an electric version in 1958.

31 One means of propelling craft through a tunnel was by shafting it with poles, which were used either as quants or against the tunnel wall – something that was frowned upon by most companies. More commonly, an unpowered craft was legged through; in the wider tunnels this was performed by men balanced on individual boards laid across the fore end of the boat or – as in this case at Bruce tunnel, Savernake, on the Kennet & Avon – by lying on their backs on the cargo. At some tunnels special teams of leggers were employed to do the work and were paid around one and sixpence per trip. Legging would often induce a lumbar complaint known as 'lighterman's bottom'.

32 With the introduction of steam on to the canal system, a number of companies built steam tugs for towing trains of boats through the tunnels, so doing away with the leggers and speeding up the passage of boats. Although the tunnel tug would go into the darkness with a full head of steam, on longer tunnels it was often necessary to stoke up at least once whilst inside. The effect of the smoke and steam on the crews of the following boats can be well imagined and it was not unknown for crews to be overcome by the fumes and at least one fatality has been recorded. Tunnel tugs were first introduced at Preston Brook on the Trent & Mersey Canal in 1864 and here at Saltersford and the nearby Barnton tunnel around the same date. Due to a surveying fault, these two tunnels had an 'S' bend in them and to assist the tugs' navigation they were fitted with spring-loaded arms with two sets of wheels which ran along the tunnel walls, so making steering unnecessary. The steering aids are just visible in this picture, which is dated 1910.

33 Tunnel tugs provided a service based on a regular timetable. Boats arriving at a tunnel would have to wait for the arrival of the tug. Here in 1930, several boats are waiting at the Welton end of the Braunston tunnel on the Grand Union Canal. The two fixing points on the counter of the horsedrawn boat were known as anser pins; to these pins were fitted tunnel hooks, on to which were tied the towing ropes to the following boat. Tunnel hooks allowed a V rope to be used so that the full movement of the rudder would not be affected. Whilst the boats were being taken through the tunnel a member of the crew, usually one of the children, would walk the horse over the top, often arriving before the tug.

34 The first steam tunnel tugs to be used on the British canals were at Standedge on the Huddersfield Canal in 1824; the Regent's Canal had one at Islington tunnel in 1826, and the Grand Junction first used them at Blisworth and Braunston tunnels in 1871. One of the problems they created was the large quantity of soot which accumulated on the underside of the tunnel roof and tended to fall on to the passing boats. To overcome this the Grand Junction used a special boat for sweeping the tunnel which had hawthorn bushes piled up on it, roughly cut to the tunnel dimensions. A newer and more efficient sweeping boat was devised in 1908 and this is the model which was produced before construction started. The new boat used wire brushes, each pivoted on a trestle. The soot dropped into a well in the boat. This new tunnel sweeper was so efficient that over ten tons was collected on its first run through both the Blisworth and Braunston tunnels.

35   The Severn & Canal Carrying
Company was, for many years, one of
the principal fleet owners on the
Severn. Around 1900 it would not have
been unusual to see a mixed train of
boats such as this, consisting of a steam
tug towing the trow *Avon* and the
trow's rowing boat, as well as an empty
and a loaded narrow boat. Towing on
the Severn was a tricky business
because of the fierce currents, but the
crews of the Severn & Canal Carrying
Company were expert at it.

36   The firm of Fellows Morton &
Clayton, founded in 1837, was one of
the largest and most famous of the canal
carrying companies. In 1923 their
boatmen were in dispute with the
management over the withdrawal of a
special bonus awarded to them during
the 1914–18 War. At Braunston, near
Rugby, one of the company's
transhipment depots, the police were
called in whilst negotiations took place.
Speedwell is one of this company's
steam-powered boats. Boats from this
fleet were often referred to as 'Joshers'
after Joshua Fellows, and they operated
over a wide area, from Preston Brook in
the north to London in the south. They
were still trading at the time of
nationalisation in 1948 and sold their
fleet to the Docks & Inland Waterways
Executive. They went into voluntary
liquidation in 1949.

37   The holidaymaker is often
pleasantly surprised by the number of
canalside public houses that he finds on
his travels. These pubs, built to supply
the thirsty needs of the boatmen, have a
very definite character of their own.
Most have names closely linked with
the waterways – The Bridge, The
Navigation, The Boatman, The Boat,
The Longboat, The Big Lock and so
on. Pubs also had stables and were often
the boatman's overnight staging post;
they were his only place of recreation
and enabled him to get away from the
cramped confines of his cabin. Now that
commercial traffic has almost
disappeared, many have turned their
backs on the canal and obtain the
majority of their trade from the nearby
roads. Some, such as The Bull and
Butcher at Napton, have had to close.
This is the Navigation Inn at
Castlethorpe Wharf near Cosgrove on
the Grand Junction, photographed in
1913; the crew of the coal cart are
slaking their thirst in the doorway.
Possibly the coal is for a steam roller
working on road reconstruction
connected with the bridge repairs.

38 Severe frost seriously hampered working boatmen. Besides making the locks and other structures dangerously slippery to work on, continuing frost will freeze over a canal. Boats could push through thin ice, but once it thickened the horse could not pull the boat through. Thick ice would also damage the hulls of wooden boats, and lumps of broken ice would get forced behind the lock gates, making them difficult or even impossible to open. A self-employed boatman had little working capital, and any days lost through being iced up were a sore blow to his trading. Families often had to spend all their saved up 'docking money' to survive really hard winters. It was unusual for rivers to freeze over and stop the traffic, but these barges were immobilised in the Trent near Gainsborough.

39 Canals can easily freeze due to the comparatively still water. This picture shows a horse-drawn ice breaker at Bridge 77, Linford Wharf on the Grand Junction canal about 1900. The crew stood on the platform of the boat, holding on to the central bar which ran fore and aft. Whilst a team of horses strained to pull the boat through the ice, the men rocked the craft from side to side, so breaking up a wide channel in the ice. It is thought that in this case the horses were not boat horses, but hired from local farmers during the spell of severe weather. There are no less than fourteen horses in the photograph.

40 During the great freeze of 1895, many boatmen were out of work as the canals had been closed for nearly seven weeks. Under the auspices of the Worcester branch of the Incorporated Boatmens Friend Society, a group of boatmen mounted a rowing skiff on a handcart and went out collecting money. No doubt the lump of ice on the cart came from the canal itself. During the same freeze-up, a group of boatmen from Whiteheath in Birmingham cut a block of ice from the canal which measured 3ft by 3ft by 2ft 6in deep, and paraded this on a handcart through the local streets with a card hung on the side saying, 'If your hearts are not as hard and cold as this block of ice, then give us help for the out of work'. The Seaman and Boatmens Friend Society lasted until 1948 and between 1894 and 1948 published a monthly journal called *The Waterman*.

42 The Whitsun bank holiday stoppage at Buckby locks in 1911. The narrow boat *Denmark* on the right is fitted with a fore cabin, where the children would have slept. Such cabins were very cramped, with poor ventilation, but were demanded in the case of large families by the Canal Boats Act.

41 Repairs to canal structures have always been necessary. Fitting new lock gates, repairing the brickwork within the locks, major work on an aqueduct or even the demolishing of a bridge, meant that the waterway had to be completely closed. In order to minimise the inconvenience to commercial craft, these stoppages were announced in advance and where possible planned to coincide with a bank holiday. Though wharfingers and labourers might be on holiday at such time, there were still plenty of boats on the move and to them even a scheduled hold-up was a nuisance, since unlike road or rail, there was often no alternative route. It has been known for as many as 80 to 100 to be waiting for the completion of repairs and the resulting traffic jams were most impressive. This stoppage is at Buckby locks in June 1907; the boat on the left is a steam-powered boat from the Fellows Morton & Clayton fleet.

43 Overloading was probably the reason for this boat sinking at Stoke Bruerne on 2 February 1924. It was a common practice on non-powered boats to load the cargo well back against the cabin, so taking advantage of the extra freeboard which was present at the stern, due to the space occupied by the cabin. If the cargo was loose, as with the coal in this case, 'slack boards' were used to contain it above the gunwales. The three-storey warehouse in the background now houses the Waterways Museum.

44 The upper reaches of the Ellesmere Canal, now known as the Llangollen Canal, contain two of the finest aqueducts on the canal system. One is at Chirk over the Ceiriog valley and the other at Pontcysyllte over the Dee. The waterway pierces right into the hills of North Wales, and for this reason it has become the most popular canal for the modern holidaymaker, with traffic jams building up on the narrow stretches near Llangollen in high summer. Though the canal is not navigable for normal craft above the wharf at Llangollen, the waterway continues for nearly two miles to the River Dee, which is the source of water for the canal. Since 1886 this two mile-stretch has featured one or two horse-drawn trip boats, taking visitors on sightseeing visits. The craft used is a double-ended skiff-type boat with a canopy for weather protection.

45 On some canals situated in pleasant areas of the countryside, it was not unusual to find a boathouse hiring out punts and skiffs. In 1908 at Catteshall Lock at Godalming on the River Wey Navigation, there were also canoes, and by this date there was relatively little commercial traffic to disturb the holidaymakers. A similar and equally famous boating station was to be found at Ash Vale on the nearby Basingstoke Canal.

46 On 18 February 1907, an accident befell the Westley Brothers & Clarke horseboat *Harold* at Heyford Bridge on the Grand Junction Canal, when it was carrying a cargo of wheat and empty sacks. Details in the records kept by the section engineer read as follows: 'Driving two cross stanks at bridges 32 and 33, opening towpath and bank at old flood paddles, letting water out and emptying and raising boat. Labour £15.11.4d; Time 28 hours. Traffic was delayed forty one hours and there were eighty-two boats detained'. The faithful dog has refused to leave his boat.

47 A new use for a then disused canal! The Thornycroft factory at Basingstoke has long been famous for making lorries for the British Army. Late in the 1920s, they are seen testing one of their gun tractors, complete with limber and gun, using the Basingstoke Canal as a natural hazard. In the area around Aldershot, the Army regularly used the canal for training purposes, building many makeshift rafts and pontoons for simulated river crossings as well as temporary bridges.

48 In 1913 the *Daily Mail* offered a prize of £5,000 to the first pilot to fly round Great Britain in seventy-two hours without alighting on land. For this challenge S. F. Cody, a famous pioneer aviator, developed special floats to be affixed to his normally land-based plane, which was powered by 120hp Austro Daimler engine. Jack Harmsworth, of the Ash Vale boatyard on the Basingstoke Canal, built the large central float and the wing pontoons. Cody carted the assembled airframe to Eelmore Flash, added the floats and launched it to test buoyancy and balance. The test proved satisfactory. Prior to this, Cody had tried the central float alone, by loading it with passengers and towing it with a motor boat between Great Bottom Flash and Mytchett Lake. It proved stable and was fitted to the plane. Cody was killed when his plane crashed a week or so before the big race.

49   During the General Strike of 1926, there was no traffic on the waters of the Rochdale Canal and no unloading at the wharves although on some canals boats were manned by volunteers. An enterprising general manager of the Rochdale Canal Company immediately made available their Dale Street wharf in Manchester for use as a car park. With all public transport at a standstill, many people came to work by car and eighty-seven of them took advantage of this extra car parking space on the first day of the strike on 4 May. No record was kept of the actual parking fee! Many makes of car are represented here, including a 'bullnosed' Morris, Daimler, Humber, Bean, AC and Austin. Incidentally this wharf is now a permanent car park.

50   Each section of a canal had its own
maintenance workshops and this one at
Ellesmere Port is probably typical.
Many tradesmen – masons, carpenters,
blacksmiths and painters – were
employed as well as labourers. This
photograph taken in 1892 shows
tinsmiths at work. Before any repair
work could have been carried out on the
canal itself, replacement parts such as
lock gates, new paddles or parts for the
company's dredger would have to be
fabricated. Each workshop would run a
fleet of maintenance boats to enable
the staff to reach the point on the canal
requiring the repairs and to transport
the bulky items such as lock gates,
bricks and mortar. Although the motor
lorry has made this easier, many parts
of the canal system are still completely
inaccessible by road.

51   In 1900 major repairs were
necessary at Nip Square lock, Walsden,
near the eastern summit of the
Rochdale Canal. These involved the
draining of the pound below the lock
and repairs to the lock walls; damage
caused by the continual buffeting of
canal craft can be seen on the right, at
the entry to the lock. Heavy lifting
equipment had to be brought in to
remove the large gates, but unlike many
of their competitors, the Rochdale
Company employed a major mechanical
aid in the form of a steam crane to help
with the task.

52   This unusual view of a staircase of
locks was taken looking up the flight at
Watford in Northamptonshire during
rebuilding in 1901–2. Four of the locks
in this seven-lock flight are in a
staircase. One lock chamber leads
directly into the next, the top gate of
one being the bottom gate of the next.
On this particular staircase water
emptied from the locks went into side
ponds to be used again, thus saving
water. The two culverts, on either side
of the bottom lock, took the water into
the pound below.

53   Another view of the rebuilding of
Watford locks in 1901–2, shortly after
the opening of the Foxton inclined
plane on the Grand Junction Canal.
Traffic was only stopped for two
months; 160 men were employed
working in shifts night and day. The
total cost was under £6,000. The
waterborne portable steam engine
driving a pump in the foreground kept
water out of the lock chambers.

54

55

54 Most locks in Britain were constructed with vertical brick or stone walls, but turf-sided locks were used on the River Wey Navigation until 1966, and can still be seen on the Kennet section of the Kennet & Avon Canal. Here Paper Court Lock on the River Wey is under heavy repair in May 1907 and much supporting timberwork is required. This type of lock was normally found on river navigations, where the heavy loss of water through the turf sides did not matter. Its main virtue was cheapness of construction, although the photograph gives the impression that such locks were expensive to repair.

55 As an aid to draining long lengths of a canal for maintenance purposes, a 'fan' boat was used. The boat was first sunk in the narrows of a bridgehole and water was lifted by the steam driven fan, via open bow doors into the adjoining canal pound. The first record of such a boat being built was in 1898; the last was used in 1938.

56 Good pumps are a primary requirement of the canal maintenance crews. Originally they were hand operated and took the form of the Archimedes screw pump shown here. The pump would be lowered into the area to be drained and, by turning the handle, water would be transferred up the screw and out of the flooded area – a slow and laborious process which was still being used on some canals as late as 1910. Slowly, however, more efficient hand pumps and latterly steam and motor driven pumps have been employed.

57 The method used for pile driving, an important part of any waterside construction, can be clearly seen here. The Thames Conservancy workmen were re-building Days Lock in 1880, which required a total stoppage on the navigation. The land-based portable steam engine would have served various purposes, including the powering of both the sawbench for all the necessary woodwork and the pumps.

57

58

58 With canals running through many miles of cutting, it was not unusual to have serious landslips. A landslip near the north end of Blisworth tunnel in 1903 took many months to repair. Moored opposite the workmen is a loaded horse-drawn boat awaiting the arrival of the tug to take it through the tunnel.

59 Major repairs inside tunnels, as in this operation inside Blisworth tunnel in April 1910, necessitated a lengthy stoppage of canal traffic. Repairs to the walling required a dam or stank to be erected; pumping was done by hand as a steam pump could not be worked in the confines of the tunnel, and suitable lighting had to be provided. It is interesting to see that the floor of the tunnel was brick-lined; this was necessary at Blisworth because of water pressure in fissures below the tunnel. In 1849 there was a partial collapse of the invert due to this pressure. A 'family group' photograph in the Waterways Museum taken inside the tunnel identifies all the workmen involved.

60

61

60 Dredging is one of the most important aspects of canal maintenance, particularly in more recent years with the coming of powered craft since their wash tends to break down the banks and deposit the silt into the channel. In early days this work was accomplished by a spoon dredger, a converted narrow boat with a small derrick and an arm to which a large iron scoop was fitted. Later came the steam dredger, such as *Waterway*, the first floating Grafton steam dredger, which was delivered to the Grand Junction Canal Company in 1896. Though the dredger is mechanised, the dredgings were taken away by a horse-drawn boat. A similar steam dredger was recently restored by the Kennet & Avon Trust, and is now helping with restoration on the Basingstoke Canal.

61 In 1910 the Grand Junction Canal Company were using a disused brickworks at Braunston as a tip for canal dredgings. Suitable tips were hard to find as they had to have good access for a crane and walls had to be built to stop the dredgings flowing back into the water. After two years or so, the land would be suitable for re-use for agricultural purposes. It was not unusual for them to be as far as five miles from the place of dredging, hence the apparent slow progress of this type of canal maintenance and the heavy cost of spoil removal. A full dredging programme demanded at least three mud hoppers, one loading, one in transit and one unloading at the dumping ground.

62 The most serious hazard on any canal is the breaching of its banks. Sometimes after heavy rain when the sluices and weirs cannot cope with the extra flow of water, a weakness in the bank is found and the canal bursts. This burst at Marbury on the Trent & Mersey Canal on 21 July 1907, shows the devastation which could be caused. Two narrow boats are caught in the burst whilst two others are stranded in the background. One of the latter is interesting as it appears to be an early example of the conversion of a narrow boat for cruising or use as a houseboat. The cause of the breach was brine pumping in the Northwich area which proved very harmful to the Trent & Mersey and Weaver Navigations, causing so much subsidence that a new stretch of canal has had to be built in recent years to bypass a section threatened by subsidence.

63 This photograph vividly if gruesomely illustrates the problems of a canal company when fences become damaged and stock strays on to the canal bank. As with the railways, the canal company is responsible for the upkeep of its hedges and fencing and part of the duties of the lengthsmen or even a lock-keeper would be hedging and ditching. Here a horse has strayed and fallen into a culvert and a set of sheerlegs has to be employed for its retrieval. Sheerlegs were more commonly used for lifting out old lock gates and dropping in new ones.

64 Steam-powered beam engines were the most common form of pump installed for canal use in the early years. Usually housed in tall, narrow buildings with an accompanying Victorian chimney, they were a familiar sight to the working boatmen. After the new pumping station had been built at Braunston, the old buildings were demolished and the chimney felled by underpinning it with wooden props which were then set on fire. Another photograph in this series shows the mighty beam of the engine being pulled through the top of the enginehouse on the right.

64

65 The fortunes of the Basingstoke Canal were never very good and as a result of lack of trade, little money was available for maintenance. This is clearly illustrated in this picture dated 1910, of Malthouse Bridge at Crookham. The bridge has already been repaired once and again the brickwork is cracking, and the banks are slowly crumbling into the water. De Salis reported in *Bradshaw* in 1904 'of late the towing-path has become much overgrown in places, thereby causing horse towing to be very difficult'.

65

CRAFT

66　The simplest form of craft used on the inland waterways was the tub boat. On the Shrewsbury Canal these measured 19ft 9in long by 6ft 2in wide. They are obviously nothing but crude floating boxes, which could carry up to five tons each; the cargo here is coal. Trains of up to twenty tub boats were common, pulled by a single horse. They were steered by a man walking along the towpath, keeping the first boat in mid-channel by means of a pole. Tub boats were chosen in preference to other craft as they were small, flexible, demanded little water and only a narrow waterway, and could easily be handled by inclined planes. On the Bude Canal, the tub boats had wheels permanently affixed to them so that they could be railed straight onto the inclined planes, thus doing away with the cradle.

67 Although 'narrow boat' is the usual term for the traditional craft of the canals, they were sometimes referred to as 'monkey boats' in the London area and 'long boats' in the Severn district. William Ward's horse-drawn boat *Fanny*, seen here at Oxford in 1900 having just unloaded a cargo of coal, is probably about to return empty to the Midlands coalfields. The wooden planking construction is unusual, being six planks deep instead of the average five. The very large rudder or helm is necessary in order to have a good purchase on the water, even when it is hardly immersed as in this picture, and the long tiller helps provide the required leverage. Usually when a boat is moored the tiller is taken out and turned up to give more headroom in the after area. The small size of the living quarters makes it hard to believe that large families were brought up in such cabins.

68 The Grand Junction Canal Carrying Department started steam narrow boat working in 1864. In 1876 they sold their fleet, some of which was acquired by Fellows, Morton & Clayton, a carrying company which built their own boats in yards at Uxbridge and Saltley. Although faster than horse-drawn boats, they did have some disadvantages: the engineroom took up the space previously occupied by some ten tons of cargo and they also needed a crew of four people. The crews of the steamers were often referred to as 'greasy wheelers' and 'greasy ockers', deriving from the cargoes of soap the boats often carried and the fact that the carrying company's headquarters were at Ocker Hill in Birmingham. This photograph is dated 1911. The lady with the large hat is Mrs Wenlock who, when visiting the Waterways Museum at Stoke Bruerne some years ago, recognised herself in this photograph, the original of which is on display there.

69 The north of England, like the east and south, had virtually no narrow canals. The narrow boat was therefore the exception and various kinds of wide boat, such as this one, were the normal form of transport. The Leeds & Liverpool Canal (the only trans-Pennine waterway still open) developed its own type of craft derived from the keels of Yorkshire and the flats of Lancashire. Most of these craft are square-sterned, giving ample scope for decoration, which was more closely connected with that of local coasting craft than the canal narrow boat. The barrel on the deck is the drinking water supply and the small kennel at the stern is the ventilator for the cabin below. This boat is loading coal near Wigan in 1890 and it is interesting to note the use of female labour.

67

70 A loaded horse-drawn boat could carry as much as twenty-five tons of cargo and it is amazing to see that two such boats could be pulled by a single horse. This was common practice on the wider waterways such as the Grand Junction or the Bridgewater and in Birmingham, but on the narrow canals it was a rare sight. The towing mast is mounted about one third of the way back from the fore end, well clear of the top planks. The bank erosion on the Grand Junction near Watford in 1928 is very apparent.

71 Some boatmen favoured mules instead of horses, mainly because of their capacity for hard work and their staying power. These mules are pulling two empty boats which have been tied side by side, 'breasted up' to use the boatman's term, since they are on a stretch of the Grand Junction Canal with many wide locks where they could enter alongside each other. Some boatmen would keep the same horse or mule for their entire lifetime, while others were born horse-traders and would be continually changing their animals.

72 By the 1920s traffic on the Stroudwater Navigation, connecting Stroud with the River Severn, had dwindled and despite a slight revival in the next decade all trade ceased in 1941. Traders on this canal were some of the last to employ donkeys (always referred to as 'hanimals') and in this case three are needed to pull the heavily loaded boat on a waterway which was likely to be shallow in places due to lack of maintenance. Obviously a good width of towpath was required for a team such as this. The boat is a Stroud barge, demasted for use on the Navigation, which formerly would have traded under square sail across the Severn estuary to load coal at Bullo Pill in the Forest of Dean. Fully loaded it had a capacity of 50 tons.

73 The Grand Western Canal was a project to link the Bristol and English channels. The eleven-mile Canal for barges between Tiverton and Lowdwells was the only part completed in 1814, but was isolated from any other water. With the failure of the channel-linking project, the Grand Western was extended to Taunton by means of a tub boat canal including one inclined plane and seven lifts. The tub boat portion was closed in 1867 but roadstone traffic continued on the remaining section well into the 1920s. De Salis visited the Canal at the turn of this century and wrote the following account in Bradshaw. 'At the present time there are apparently only two boats on the canal which are engaged in the roadstone traffic. They work chained together fore and aft – the foremost one which has a pointed bow, carrying about 8 tons, and the after one, which is a box boat, carrying about 10 tons; each draws, when loaded, about 1ft 8in of water.'

74 Various types of motive power were tried by canal carriers. However, it was not until 1910 that the first semi-diesel, a Bolinder, appeared in a hull in this country and then it was in a Thames barge. The single-cylinder Bolinder was at first the most popular, but later gave way to Gardner semi-diesels and later full diesels, mainly Nationals. Motor boats could easily tow an unpowered boat, always referred to as the 'butty' and many horse boats took on a new lease of life as towed butties. When operating on a long stretch of lock-free water or on rivers, they were often towed as far as 70ft behind the motor boat, the distance being controlled by the butty's steerer, as the tow line ran through running blocks to a fixing point immediately in front of him on the cabin top.

75

76

75   The Monmouthshire Canal started
at Newport and had two branches, each
eleven miles long, one to Crumlin and
the other to the Brecknock &
Abergavenny Canal at Pontymoile near
Pontypool. The Canal Company turned
itself into a railway company in 1845
and was bought up by the Great
Western Railway in 1880. Because of
the unusual dimensions of the locks on
these canals, boats were usually 60ft by
8ft 6in and could carry 20 tons on a
draught of 2ft 9in. This boat is
negotiating a most uncommon type of
double lifting bridge carrying a railway
siding on a very sharp curve near Mill
Street in Newport. The last commercial
traffic at Newport was loaded in 1915,
around the date of this photograph.

76   At one time there were many
boatyards along the canals, each making
boats roughly to the same design but
with its own individual features. Bushell
Brothers' yard at Tring in 1908 would
have been typical. Each yard would
have its own painter, and the expert
could tell from which yard a boat came
by the shape of the hull and the style of
the roses, castles and other decoration.
It was usually necessary for a boat to be
docked for repair each year, which was
the time the boat family would take
their holiday – if they could afford it. It
was on these holidays that they
collected many of the souvenirs, such as
lace-edged plates, which adorned their
cabins.

77   The directors of most of the canal
companies had their own boats in which
they could travel to inspect their
waterways. This is the inspection craft
of the Grand Junction Canal Company.
Sitting in the well at the fore end is the
bearded Rudolph Fane de Salis, who
was chairman of the Grand Junction
from 1914 until 1928, the end of that
company's independent life. In the
background of the picture is an early
experimental motor boat, most
unusually fitted with wheel steering.
Little is known about *Mermaid*, but one
of the first motor canal boats called
Progress was launched on the Worcester
& Birmingham Canal on 24 August
1907. This craft was capable of 5 to 6
miles an hour, and it consumed half a
gallon of paraffin (at 5d a gallon) per
mile. *Progress* was also wheel steered,
but these craft were the exceptions.

78   Narrow boats regularly used the
Rivers Severn, Trent and Thames and
if it was not possible for them to be
towed by horses they were usually
pulled by tugs. In general these boats
were not designed for sailing, which was
obviously quite impossible on the type
of waterway for which they were built.
However, this photograph was taken in
March 1888 and clearly shows a pair of
narrow boats 'breasted up' under sail
entering a lock on the Mersey and
Irwell Navigation. One of the crew is
assisting by shafting from the top of the
cabin. Narrow boats working off the
Chesterfield Canal were sailed regularly
on the River Trent; they set a square
sail and had windlasses fore and aft to
handle it and the mast.

79   The Humber sailing keel was one
of the principal craft trading on the
rivers of Yorkshire well into the present
century. The last to trade under sail was
the *Nar*, which came out of service in
1949. The keels had bluff bows and
stern which enabled as much cargo as
possible to be carried within the limits
of length dictated by lock sizes on
connecting waterways. Here in 1900 one
is being shafted out of Keadby lock
from the River Trent on to the
Stainforth & Keadby Canal. For use on
canals a short iron tiller was fitted to
avoid fouling lock sides. As trading
under sail came to an end, the masts
and rigging were removed and the hulls
used purely as dumb barges.

80   Apart from tunnel tugs, steam tugs only operated over a few lengths of the canal system, although a fleet of them was worked by the Bridgewater Canal. These tugs towed flats from the top of Runcorn locks to Preston Brook and Manchester. Larger tugs, also owned by the Bridgewater concern, brought flats from Liverpool Docks to Runcorn. Sometimes tugs were used for towing horse-drawn narrow boats on the Manchester Ship Canal. This picture, taken in 1906, shows Waterloo Bridge at Runcorn, hard by the top of the flight of locks leading down to the Ship Canal. The two white bands on the chimney were the funnel colours of the Bridgewater Trustees and later of the Manchester Ship Canal Company. It is reputed that the coming of the steam tug to the Bridgewater Canal was speeded up by an epidemic amongst the company's horses, which resulted in the death of over two hundred animals. The tugs came in during the years 1875 and 1876; they eventually numbered twenty-seven in all and were named after local places.

81   The Aire & Calder Navigation from Goole to Leeds and Wakefield has always been one of the most progressive of our inland waterways. One of its special features was the long trains of compartment or 'tom pudding' boats, often up to forty at one time. Designed and patented by W. H. Bartholomew, engineer of the Aire & Calder, they started operating in 1865. The compartment boat was an iron box 21ft long, 15ft wide and 8ft deep, capable of holding thirty-five tons. Designed for coal carrying, these trains were towed by tugs and the front compartment had a dummy bow attached to it to improve the propeller flow and steerage. These craft were usually loaded from gravity chutes at the side of the waterway, the coal being transported to the staithes in railway trucks. At Stanley Ferry Basin, some 'tom puddings' were floated on to a twelve-wheeled bogie-rail carriage and taken directly to St John's Colliery at Normanton for loading. The locomotive was able to pull the compartment up the slip and out of the water by means of a snatchblock and it checked the descent of loaded compartments by using steam in its cylinders as a brake. This practice ceased in 1941 when mining subsidence damaged the railway track. At the time of writing, British Waterways still have a fleet of three tugs and 430 'tom puddings' in regular use.

82   When the 'tom pudding' boats reached the port of Goole, they were lifted bodily by a hydraulically operated hoist and tipped into the holds of waiting sea-going colliers, the first of which was installed in 1868. It was the foresight of Bartholomew through the compartment boat system (on which he got a royalty of $\frac{1}{2}$d per ton, incidentally) and the direct loading into ships, which brought the port of Goole into prominence. Of the five hoists that he built, two are still in existence and one in regular use.

83 The waterways of East Anglia had their own type of trading craft, of which the Norfolk wherry is the most famous. The size of the wherry varied tremendously, depending upon where it traded in the river and broads system. The wherry had a fore and aft gaff-rigged mainsail with no boom, no shrouds, one halliard and a very efficient halliard winch which enabled the boat to be handled by a crew of only two. The River Bure, one of the longest in the East Anglian waterways, was navigable at one time from Great Yarmouth to Aylsham, forty miles inland. However, a great flood in 1912 damaged the top nine miles which were never repaired. Here, a wherry is unloading at Coltishall Mill, some thirty miles inland, in 1902.

84 The Severn trow was that river's own brand of sailing vessel and some say its history can be traced to the fifteenth century. It was built in varying sizes and worked on both sides of the estuary and up some of the neighbouring canals. Some of the smaller trows were even built on the Kennet & Avon Canal as far away as Aldermaston and Honeystreet. The Danks family, who were proprietors of what became known as the Severn & Canal Carrying Company, introduced iron-hulled trows, and between 1843 and 1876 ten were built for river and coastal work. The Danks' iron trows were named after rivers; this one, the *Avon*, was built in 1858 and later modified to bring her total tonnage to 65 tons. She eventually became a towed lighter and was finally broken up in 1924. This photograph must have been taken before the 1880s, when this depot at Redcliff Backs in Bristol was demolished.

85   Coal was the major cargo on the inland canal system from first to last. Even though the railway and later the roads took the traffic from the canals, it was the conversion to oil firing which finally put paid to regular runs by the narrow boats. John Dickinson, the papermakers in Hertfordshire, ran a fleet of narrow boats for many years. Some of these boats were used to bring coal to the factories from the Midland coalfields, whilst other boats were modified to carry rolls of paper away from the mills. These boatmen were often called 'paper dashers'. One of their regular runs was from Paddington Basin to Croxley carrying rags, waste and shavings and then back to Paddington with the paper rolls. This service was operated at first by Fellows, Morton & Clayton with steamers and butty and later with motor boats like these owned by John Dickinson; the service survived until the 1950s.

86

87

86　Unfortunately this basin at Kidderminster has now been filled in, though some of the wharf buildings remain. In the early 1920s, coal was still the main trade to the wharf and the horse and cart was still much in evidence as the principal means of distribution. Coal, in and around the Midland coalfields, was transported in narrow boats with small cabins, the craft being known as 'joey' or 'day' boats since the crews returned home each night instead of living on the boats. The small cabin or lack of it also allowed more cargo to be carried. The boat in the foreground is a long distance boat which has been converted to a day boat, whilst the one against the wharf is a true day boat.

87　Coal was required not only by industry but also the private consumer. Private coalyards such as this one at Heyfields Wharf, Oxford in 1890, were commonplace and trade with them was the domain of the 'number ones' or self-employed boatmen. The Oxford Canal in particular saw trade of this sort until a late date, some 'number ones' employing horse power into the 1950s. Note the complete absence of any form of mechanical unloading aids; a twenty-ton cargo of coal would take a two-man crew about half a day to unload – the work becoming progressively harder as they dug deeper into the hold.

88　It should not be thought that railways were always a threat to all canals. Often tramways and railways were laid from inaccessible mines and quarries down to a canal where the goods could be easily carried away and in some such cases this was more satisfactory than laying a new railway line. Froghall, on the seventeen-mile long Caldon branch of the Trent & Mersey Canal, was a good example. This canal was at one time owned by the North Staffordshire Railway who also owned limestone quarries at Caldon Low, which were connected to the Froghall transhipment basin by a $3\frac{1}{2}$-mile tramway. This tramway, cable-hauled since 1849 was closed in 1920. This photograph, taken in 1910, shows the bustle of what was a very busy canal wharf as the horse-drawn narrow boat *Dorothy* is loaded.

89 Milk was regularly carried on the
Shropshire Union Canal (about 120 churns
per boat) from the Shropshire farmers to
Cadbury's factory at Bridge 45 near
Knighton. These boats are about to
enter the bottom of the five locks at
Tyrley, some six miles from their
destination. Cadburys first started
transporting this cargo in their own
boats in 1912 and they also carried loads
of 'crumb' (dry powder of cocoa, sugar
and evaporated milk) from Knighton to
their other canalside factory at
Bournville, Birmingham, loading back
'mass' (pure ground cocoa). Both
factories were equipped with covered
unloading wharves and are in good
repair today, although now used as
garages for motor lorries.

90 Two different ways of covering the
cargoes are shown here, as two horse-
drawn boats leave Iron Bridge lock in
1928. The boat on the left has its cargo
sheeted up in the traditional style over
the central gangplank, which allows the
crew to move from the fore end to the
stern of the boat over the cargo. The
other boat was probably carrying a
cargo of boxes which, because of their
width in the hold, did not allow the
traditional method of sheeting.
A. Harvey Taylor was an Aylesbury-based
carrying firm which was acquired by the
Samuel Barlow Coal Co Ltd in the
1950s. The bicycle on the top of the
cargo is typical, as a member of the
crew would cycle ahead on a heavily
locked section to set the locks in
advance of the approaching boats. The
frontispiece shows the same scene
twelve years earlier; compare the wear
on the bollards.

91 The main source of revenue of the canal companies came from tolls charged to the firms or individuals who operated over their waters. Tolls were worked out on the weight and type of cargo being carried, and toll offices were set up at intervals along the canal. In order to work out the amount of cargo being carried, each boat was 'gauged' at least once on each trip; gauging consisted of measuring the freeboard or 'dry inches' of the boat. When a boat was built or substantially altered in any way, it had to be measured with varying tonnages aboard and these displacements were then sent to each toll office, since it was not sufficient to believe the tonnages marked on the cargo-loading slips. Here, the Fellows Morton & Clayton steam narrow boat *Phoenix* is being gauged at Buckby toll office by a toll clerk in knickerbockers.

92 Thomas Clayton of Oldbury, founded in 1842 as general carriers, were later one of the few canal carriers to specialise in liquid cargoes in bulk. Their cargoes usually consisted of crude tar or 'gas water' (for ammonia production) and they operated between the gas works and the tar distilleries. They also offered a service carrying crude and refined petroleum from Stanlow to Oldbury and Trafford Park to Oldbury. The boats were always recognisable by the flush decks, the cargo being poured directly into the hold – the boats thereby becoming floating tanks. In this case some barrels are being carried as extra deck cargo. With the demise of town gas works, the traffic gradually dwindled and the firm gave up canal carrying in 1966, though it is still in business in liquid fuel delivery by motor lorry.

93 A partially loaded Thomas Clayton liquid-carrying horse boat pauses at Linford Wharf in Buckinghamshire. The scene is almost timeless, although the photograph is thought to have been taken around 1930. The boatman's wife is taking advantage of a delay to hang up the washing; menial chores such as washing or shopping were difficult to fit into the very long working hours of the boatmen. The 'cratch' on top of the cargo-hold is for the storage of horse feed. Many of the cargoes carried on the canals were of a perishable nature, and often the wharves would have some form of overhanging protection in case of inclement weather.

94   Since the 1890s, canals have been
the haunt of the picture postcard
photographers who have given us many
interesting scenes. In some cases,
however, the postcard publishers
cheated and this is a good example. The
scene is the Ogley flight of locks (now
closed) on the Wyrley & Essington
Canal near Lichfield in Staffordshire.
Presumably it was too dull as it stood
and so a boat with a towing horse has
been added. As this is a narrow canal
and the boat inserted is a wide one,
possibly a Yorkshire keel, it could never
have passed through the flight of locks.
The card is postmarked July 1908.

95   One of the largest industrial
concerns served by the canal system was
the Stanton ironworks near Ilkeston in
Derbyshire. Coal and limestone came
by water, and iron products, including
the famous Stanton pipes, were taken
away. Situated on the Nutbrook Canal
off the Erewash Canal, the loading
wharves were a scene of continuous
boating activity.

96 The Monmouthshire Canal in South Wales was built to serve the blast furnaces at Abercarn, Ebbw Vale, Nantyglo and Blaenavon and was fully completed in 1799. Though most of the cargo was for industry, a regular service of boats took perishable cargoes to and from market in Newport. This photograph shows the last market boat to leave Newport, being loaded on 9 January 1915; after this cargo had been delivered the crew joined the services and were posted to France. The hand-operated vertical lifting bridge behind is an unusual feature. The boats on this canal and the nearby Glamorganshire Canal did not display the highly decorative paintings of the Midlands system; the single lozenge on the bow was the main decoration, whilst some craft had a propeller-shaped design on the cabin side with the carrier's name, and there was sometimes a quartered lozenge design on the stern.

98 Where it was possible for craft from the inland canals and waterways to reach sea-going vessels, cargoes could be shipped overside, so greatly reducing the time and labour costs. These topsail schooners were engaged on the flint, felspar and pottery materials trade, some of it coming from Devon and Cornwall. This scene was taken in 1890 at Fenton Dock, Runcorn, and shows two narrow boats loading overside for onward transmission to the Potteries. The building in the background is a six-storey pottery warehouse, where crated crockery for export was stored.

97 This undated picture of the Town Wharf at Guildford on the River Wey Navigation must have been taken around the turn of the century. The brewery on the left is now the bus station, whilst the wharf is an open space and riverside walk. Just visible on the wharfside is a most unusual crane, powered by a treadwheel in which a man used to walk. Though coal was one of the principal cargoes on this waterway, there was much grain traffic to the various mills, the last boats reaching Coxes Mill at Weybridge in 1969.

9   Large ocean-going ships dominate No 9 dock in Manchester in April 1930, at the head of the Manchester Ship Canal. Wool from Australia and New Zealand could be loaded overside and taken to the mills of Yorkshire, whilst American and Egyptian cotton went to the mills of Lancashire.

100   In the days of heavy commercial traffic on the northern waterways, it was not unusual to see queues of boats waiting to unload at any one wharf. In 1905, possibly on a Sunday, ten keels or 'west country boats' loaded with grain lay at Woodside flour mills at Elland on the Calder & Hebble Navigation. These boats could have been loaded from ships in Manchester docks and come up the Rochdale Canal joining the Calder & Hebble for the three-mile run to Elland, or from Hull docks via the Aire & Calder Navigation.

101   Most of the British canal system is associated with industry of one sort or another, but some waterways were built purely to serve agricultural areas; the Bude canal being typical. The canal was completed in 1825 and its main purpose was to take sea sand from the coast to inland farmers, where it was used as a fertiliser. The horse-drawn railway was laid from the canal wharf on to the shore at Bude, to facilitate the collection of the sand, and it was then transferred from the tipper trucks to the canal tub boats for transmission inland.

102   It was important to have dry
storage space at the larger canalside
wharves. Some of the goods would be
awaiting boats, whilst newly-arrived
merchandise would be awaiting
collection or distribution. The largest
warehouses, such as this one, would be
at the docks, but on the inland
waterway system a place such as
Braunston near Rugby had large storage
areas, as it was a transhipment point for
the fast steam narrow boats coming up
from London. The cargoes would then
be taken on, often by horse-drawn
boats, to some of the lesser used
unloading points. In some places,
warehouses were not large enough to
accommodate all the loads and boats
had to wait days to unload. On the
Duke of Bridgewater's Canal the boats
were even charged for the time they
occupied waiting to be unloaded in the
basin!

103 Canal companies who ran their
own fleet of boats trading on their own
and local canals had to have a fleet of
land-based vehicles in order to collect
or deliver their cargoes to the wharves.
In July 1914, the Rochdale Canal
Company took delivery of a 40–50 hp
Leyland lorry on a month's trial, to see
if motor transport was more economical
than the horse and cart. The lack of
weather protection for the driver and
the solid tyres of the lorry were
commonplace at this time on
commercial vehicles. This lorry
operated in the Halifax and Sowerby
Bridge areas and proved to be extremely
useful. The canal company agreed to
purchase it for £700, but unfortunately
on 12 August, the day before officially
taking the lorry over, it was
requisitioned by the War Department.
The lorry was not returned after the
war and the canal company never
replaced it.

104 The traditional decoration of the narrow boat had been a notable feature of the canals for well over one hundred years, but no two historians agree as to the origin of the roses and castles on the cabin doors, or the larger castles and scenes on the side of the cabin. Through the open cabin door can be seen an oil lamp, quite definitely the only lighting in the cabin. H. R. Robertson, in *Life on the Upper Thames* published in 1875, has this to say about the narrow boat cabin . . . 'The spotless

decoration of these boats is noticeable, and evinces the pride taken in their appearance by the owners who repaint them with the gayest colours as often as they can afford to do so. On the outside of the cabins are painted two or four landscapes . . . The smartness of the cabin part of the barge is often the more striking, from the fact that the load it bears is of a very opposite character, as coal, which is perhaps the most common freight.' This boat is owned by a private trader, or 'number one', as

105 This picture displays many of the familiar trappings of the narrow boat families, in particular the two water cans, originally brightly painted, which were the only means of storing drinking water on the boats, and the bucket, or sometimes a dipper, for taking water from the canal for washing or for watering the horse. Many families had pets, such as the bird in the cage, and many had a dog which was sometimes used for poaching! Much has been written about the gay and bright side of the boat families' life, but this picture of

to-day dress. George Smith, writing in 1875 in *Our Canal Population*, gives the other side of the coin: 'Their habits are filthy and disgusting beyond conception . . . they wash their clothes – those they do wash – out of canal water, and instead of being white, or near to it, they look as if they had been drawn through a mud hole, wrung, and hung out upon the boat line to dry. . . . swearing is taught to the children before anything else. . . . the women are coarse and vulgar, and, if anything can outdo the men in resorting to obscene

106   The families that worked the boats on the inland canals were hard-working people, toiling long hours seven days a week, and naturally they were very fond of occasions which gave them a chance to dress up in their best clothes. A marriage linking two families offered a perfect occasion for celebration, as must have happened at the time of the wedding breakfast at the Britannia public house at Thrupp on the Oxford Canal in 1915. The marriage was between William Gibbins and Fanny James. The bride's sisters, Martha and Andria, make up this pair of rather nervous-looking bridesmaids. Boat people were a race apart and tended to marry amongst themselves. When a girl from 'off the land' did marry into a boating family, there was often hostility from her family. At one such wedding at Pelsall the couple had to be protected by an armed guard of boatmen.

107 E. King, often referred to as
Grandad King, was born in 1831 and
when he was eight years old was stolen
from a boat near the Fishery Inn at
Boxmoor by a Manchester boatman.
For three years he was forced to work
for this man and when he was eleven
the boat returned to the south and he
was promised a trip out to see his
parents. However, the rascally boatman
broke his promise. He took the boy's
trousers away and locked him in his
cabin, but Grandad King found a way
out and escaped home without them. At
a later date he returned to the canals as
a mate on the express flyboats. When he
was about seventy, he retired from the
waterways and died in 1907.

The Canal, Berkhamsted

108 From an early age, the children of narrow boat were expected to help with the working of the boat, and when only about three years old they were taught to steer the boat, standing on a stool to make up the height. Many of them never went to school and were often illiterate, or at best they snatched a few hours schooling when a cargo was being unloaded. Here two boat families pause at a lock at Berkhamsted in 1905.

109 Christenings were another occasion for the family to turn out in full traditional apparel. The ladies had a very full, long skirt, boots and elaborate black bonnets which came in at the time of the death of Queen Victoria, but continued to be worn for years after the period of mourning was over. The men wore corduroy trousers, waistcoats sometimes made out of moleskin, and flat caps. The children were usually dressed in a miniature version of the grown-ups' clothing. Though this photograph was taken in 1913, many boatwomen continued to wear this style of dress right up to 1939.

110 On river navigations it was necessary to have personnel responsible for the various weirs and flash locks, particularly if they were of the older style with removable paddles or sluices. Radcot Weir on the Thames was looked after in 1900 by a man called Harper, who was also a keen fisherman and naturalist.

111   On the canal system it was usual
to have one lock-keeper to look after a
flight or series of locks, his duties being
aimed at surveillance and maintenance,
rather than actually helping with the
passage of the craft. On rivers, however,
there was usually one lock-keeper per
lock and he operated the mechanism.
James Lowe had been lock-keeper at
Days Lock on the River Thames since
1870, although this picture was actually
taken in 1904. The level marking the
highest point of the floods in 1894,
shows that his job was not always such a
pleasant one.

112　William Grove, lock-keeper of
Triggs Lock on the River Wey
Navigation from 1856 to 1915, was a
familiar sight to all the barge crews.
The River Wey is navigable from
Godalming to the River Thames and it
was only in 1969 that commercial traffic
ceased to operate on this waterway. The
last owner, William Stevens,
bequeathed it to the National Trust in
1963 and the length above Guildford
was given to the Trust in 1968. The
River's heyday was before 1845 when,
together with its connections with the
Basingstoke and the Wey & Arun
Canal, 'the River Wey . . . is known to
have exceeded the most sanguine
expectations of the owners'. Traffic
declined after 1845 with the coming of
the railway.

113　Fishing has always been a popular
sport on the canals and this 1905 scene
is almost timeless; only the clothes have
changed and the loaded narrow boats
have disappeared. There are at present
some 1·1 million fishermen on the fresh
waters of this country, of which over a
quarter of a million are known to fish on
the canals. Fishing techniques have
changed little over the years, the only
changes being in tackle and bait. The
most popular fish to be found in canals
are roach, perch and bream, whilst
dace, tench, chub and carp are
regularly caught. Pike are common and
on some rivers eel-traps are still set.

114   The *Mary Ellen* belonged to a fleet of Calder & Hebble style keels, operated by the Rochdale Canal Company, and because of her restricted length could travel on all but the narrow canals of the north. With a maximum beam of 14ft 3½in, the boat was capable of carrying 41 tons, though in this case in 1899 the cargo was a happy band of holidaymakers on a works' outing. Two horses are being employed to move the boat to ensure minimum waste of time, though under normal circumstances, only one horse would have been used.

115 The boatman's livelihood depended on the horse, which was naturally regarded as part of the family. The harness was often decorated with brightly painted bobbins and many horse brasses. Great care was taken to keep the flies away from the ears by covering them with ornately crocheted head-gear. Some boatmen were so proud of their animals that they would enter them in horse fairs, where they would frequently take away the top prizes. One member of the crew was detailed to walk ahead with the horse to ensure an even pace was kept whilst travelling; in this way they must have walked over a hundred miles in an average week.

ACKNOWLEDGEMENTS

Several hundred people have helped with the collection of the material and information in this book. Credits to individuals and organisations are listed below; to them and to the many whose material has not been used, a very sincere thank you. I am most grateful to British Waterways for permission to use a number of photographs from the Waterways Museum at Stoke Bruerne and for the help that has been given me throughout by Richard Hutchings, Curator of the Museum. Edward Paget Tomlinson was kind enough to read the manuscript and made many helpful suggestions and corrections. It was Charles Hadfield's enthusiasm which led to the compiling of the book and he has kept a fatherly eye on its development. As well as providing much additional information, he was kind enough to read the manuscript and proofs. To Jane Rieman who typed many letters and Doris Draysey who typed the final manuscript – thank you.

Hertfordshire County Records Office: Frontispiece, 70, 71, 74, 85, 90, 101, 115.
Waterways Museum, Stoke Bruerne: 1, 4, 5, 7, 10, 18, 19, 22, 23, 30, 34, 36, 37, 39, 40, 41, 42, 43, 46, 52, 53, 56, 58, 59, 60, 61, 62, 63, 64, 66, 68, 77, 81, 84, 91, 104, 105, 106, 107, 109.
Science Museum: 2.
York City Library: 3.
Gloucestershire County Records Office: 6, 25.
Manchester Ship Canal Company: 8, 9, 11, 13, 14, 16, 24, 27, 50, 78, 98, 99, 102.
Manchester Public Library: 12.
Oxford City Library: 15, 29, 57, 67, 87, 110, 111.
Dudley Public Library: 17, 55, 86.
British Railways, Western Region: 31, 44, 73, 75.
Ironbridge Gorge Museum Trust: 20.
Author's Collection: 21, 28, 32, 35, 47, 80, 94, 95, 108, 113.
Charles Hadfield: 26, 88.
C. Grey: 33, 92.
Gainsborough Public Library: 38, 79.
Couzins-Powney Collection: 45, 65, 83.
Royal Aircraft Establishment, Farnborough (Crown Copyright): 48.
Rochdale Canal Company: 49, 51, 100, 103, 114.
Guildford Museum: 54, 97, 112.
John Cornwall: 69.
Mansell Collection: 72, 93.
Miss C. Bushell: 76.
Goole Public Library: 82.
Michael Norris-Hill: 89.
Newport Public Library: 96.